Anatomy & Physiology
#1
Bones Muscles and The Stuff That Connects Bones and Muscles

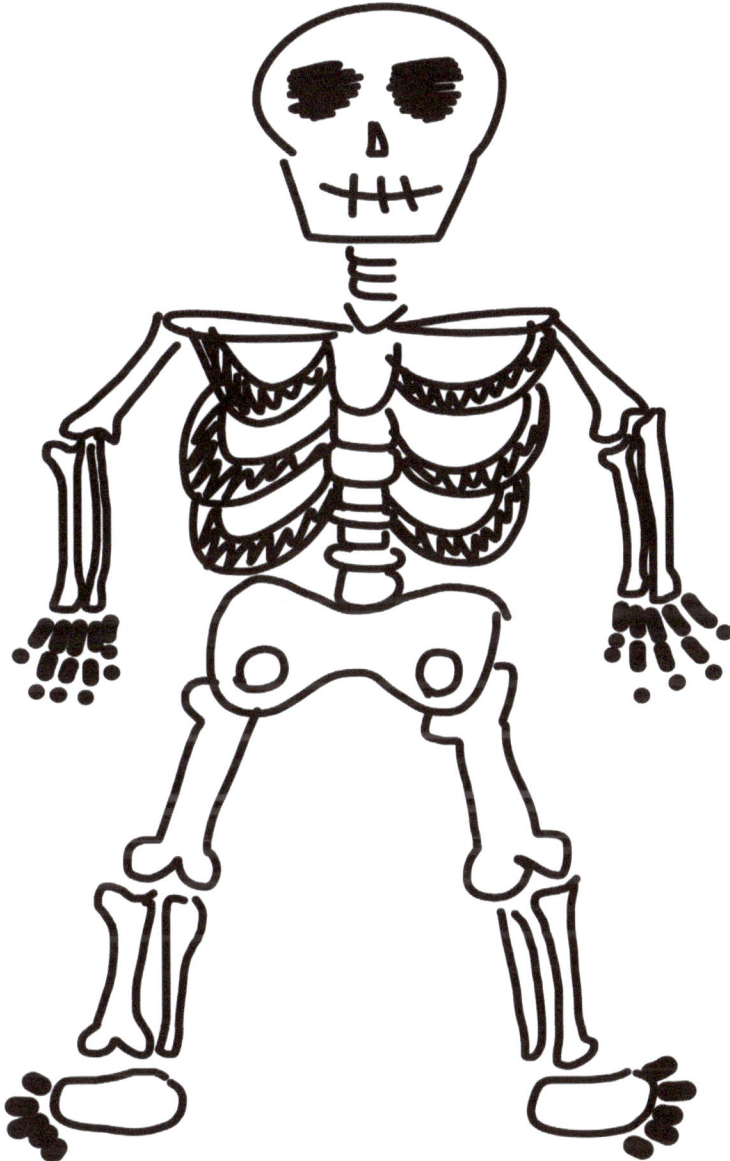

This book is dedicated to my Mom & Dad, Kim and Joel Terrazas.

Thank you so much for your wholehearted support of all of my endeavors since early childhood!

By: April Chloe Terrazas

Anatomy & Physiology PART 1: Bones, Muscles, and The Stuff That Connects Bones and Muscles.
April Chloe Terrazas, BS University of Texas at Austin.
Copyright © 2014 Crazy Brainz, LLC

Visit us on the web! www.Crazy-Brainz.com
Cover design, illustrations and text by: April Chloe Terrazas

Anatomy art competition winners:
Taj Estrada and Sydney Estrada!

Congratulations!
We LOVE your portrayal of
human anatomy!

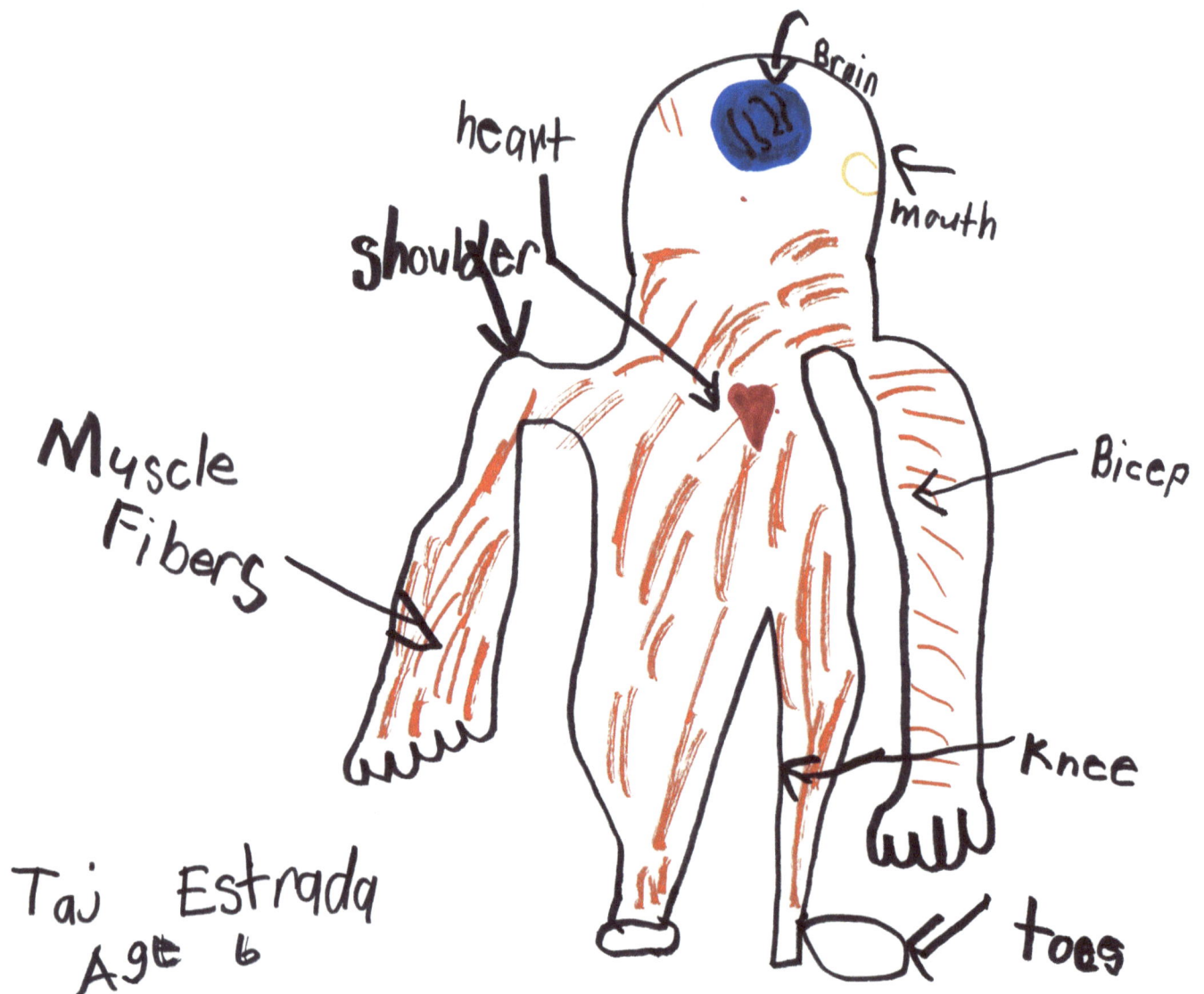

Brain

mouth

heart

shoulder

Muscle
Fibers

Bicep

Knee

toes

Taj Estrada
Age 6

The Inside of a Bone

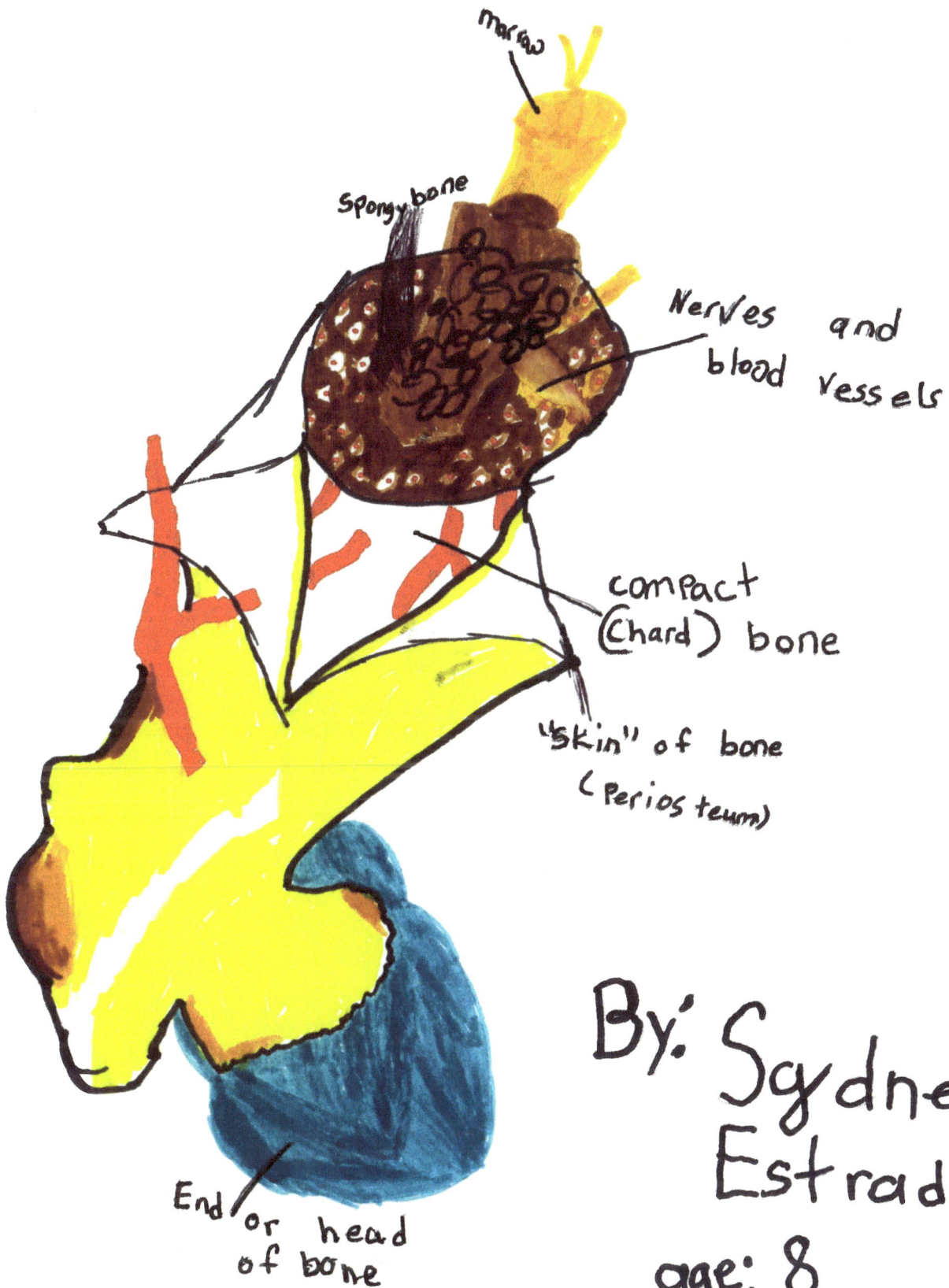

marrow

Spongy bone

Nerves and blood vessels

compact (hard) bone

"skin" of bone (Periosteum)

End or head of bone

By: Sydney Estrada

age: 8

Anatomy

Sound it Out
1. UH
2. NA
3. TO
4. ME

Physiology

Sound it Out
1. FIZ
2. E
3. OL
4. O
5. JEE

Anatomy is the structure of a living organism *(how it looks)*. For example, the hand is small with fingers on it.

Physiology is the function of a living organism, *(how it works)*. For example, the hand moves to write or play an instrument.

What is the anatomy
of your face?

How does it look?

Eyes
Nose
Mouth

What is the physiology
of your face?

How does it work?

See
Smell
Talk
Eat

Bone

Skeleton

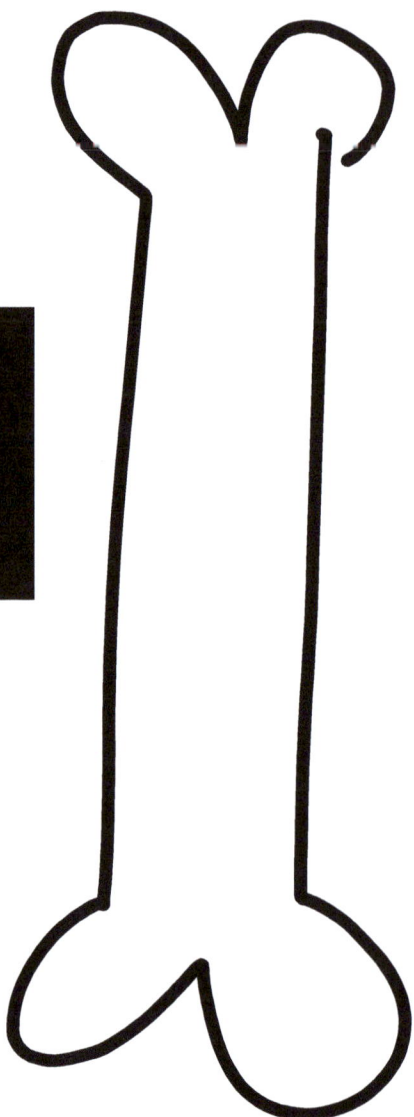

Bones are hard tissue that make up the skeleton. Without bones, we would be like a ball of jelly!

Bones are alive and can grow and change shape.

A human adult has 206 bones!

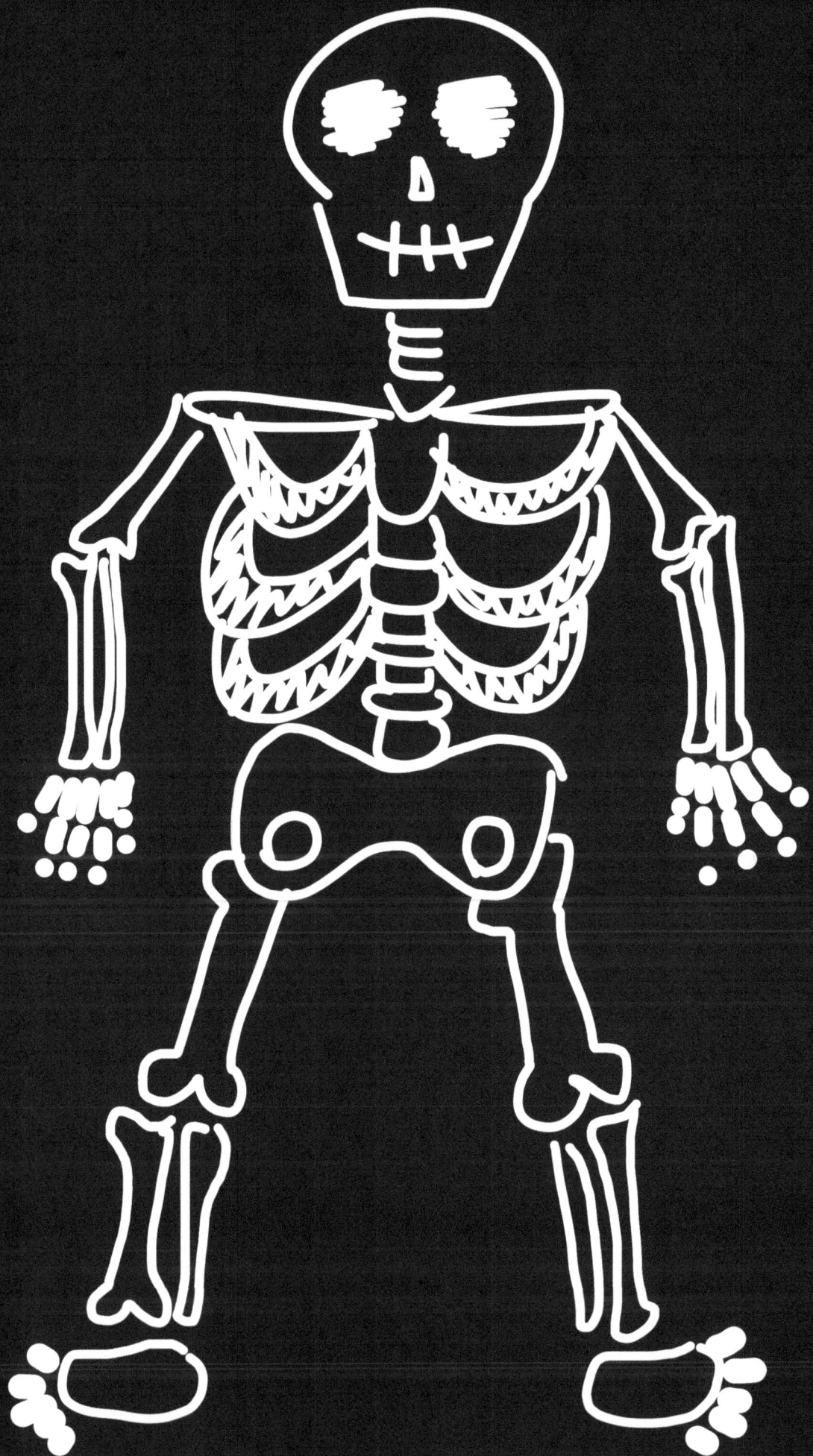

This is a <u>long bone</u>.
<u>Long bones</u> are in your legs,
arms and even your fingers!

The epiphysis
is the end of a
<u>long bone</u>.

The diaphysis
is the shaft of a
<u>long bone</u>.

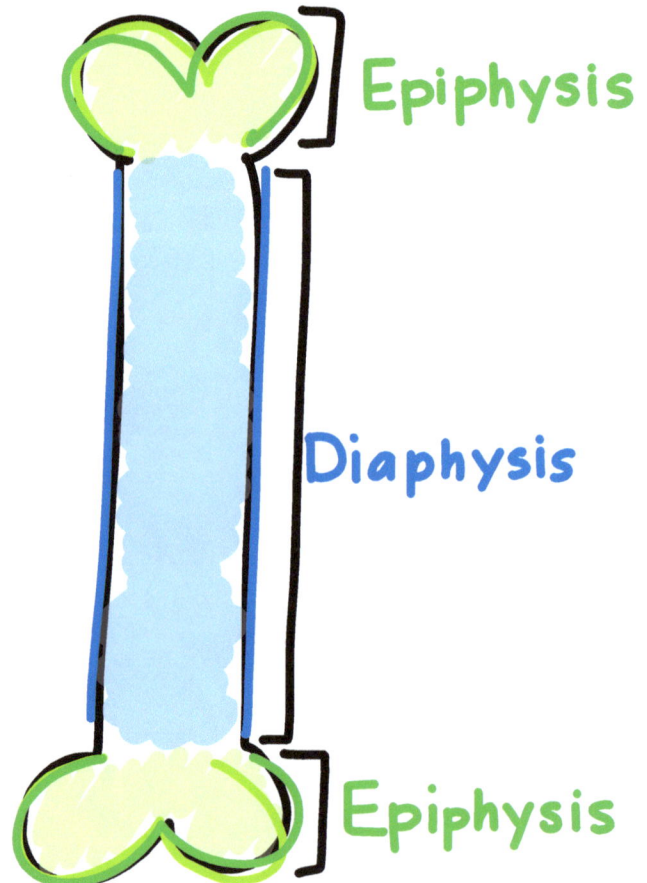

Epiphysis

Diaphysis

Epiphysis

Epiphysis

Sound it Out

1. E
2. PIF
3. EH
4. SUS

Diaphysis

Sound it Out

1. DI
2. AF
3. EH
4. SUS

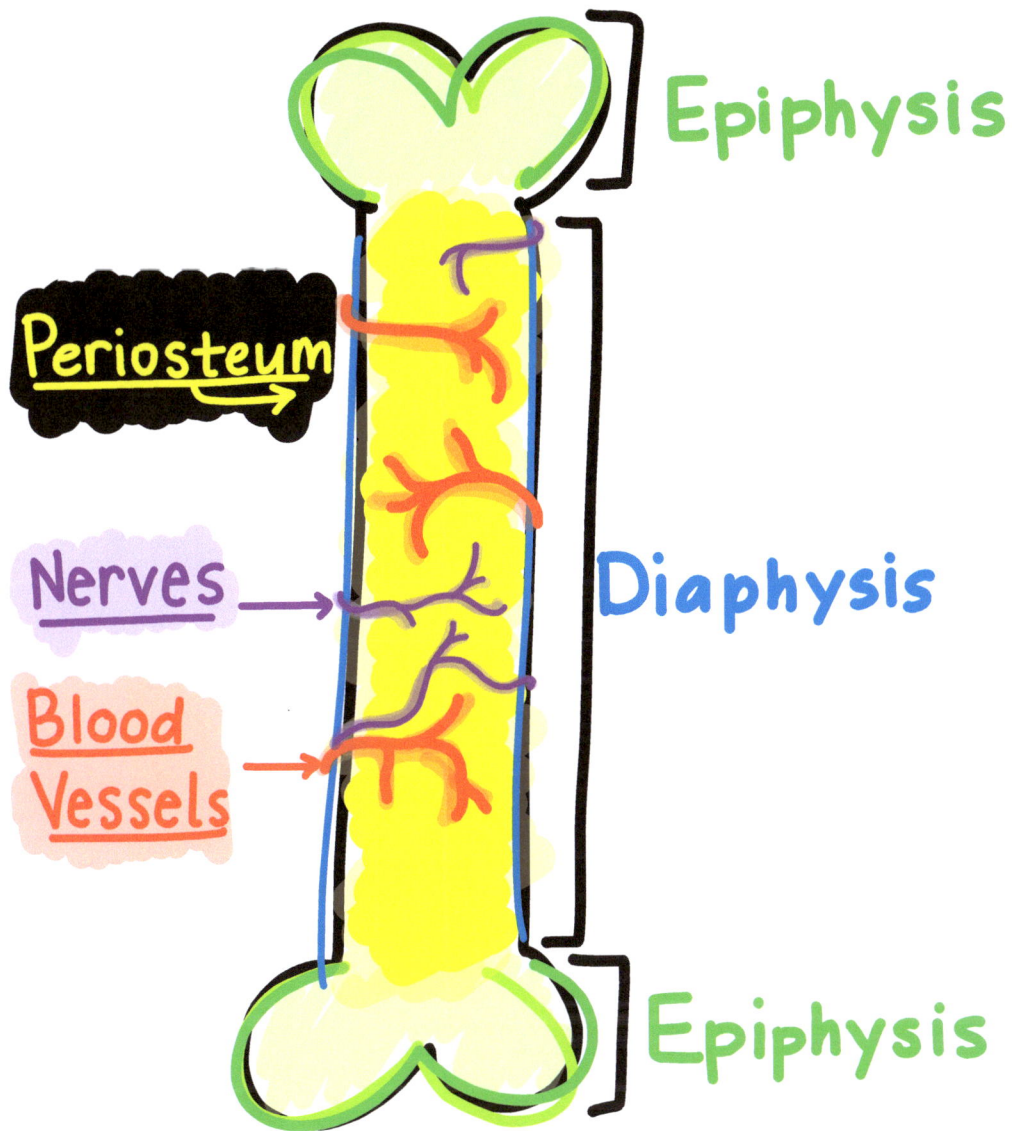

Epiphysis

Periosteum

Nerves

Blood Vessels

Diaphysis

Epiphysis

Periosteum is a membrane that covers the outer surface of bones and has **nerves** and **blood vessels**.

Periosteum

Sound it Out

1. PER
2. EE
3. OS
4. TEE
5. UM

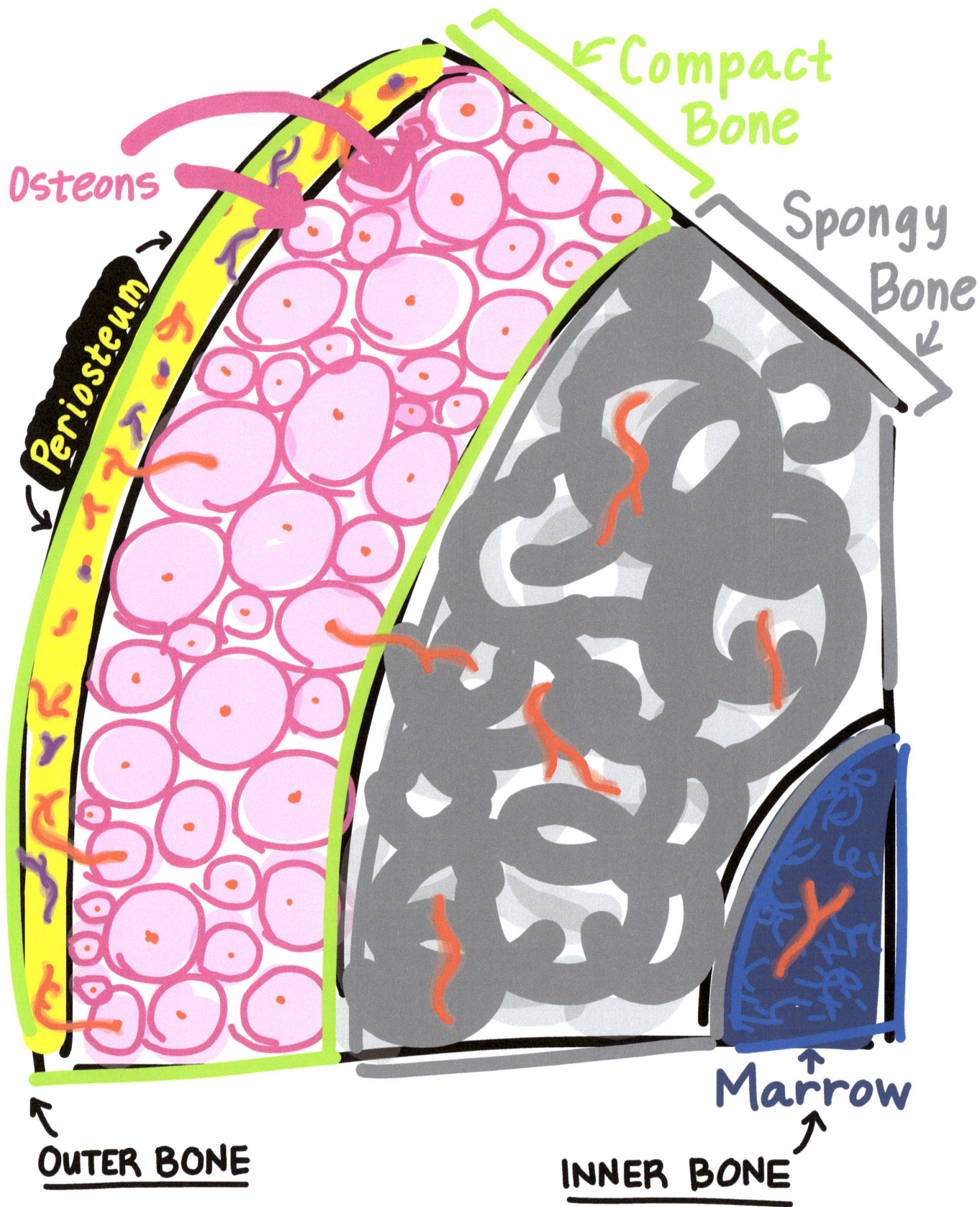

Osteons

Periosteum

Compact Bone

Spongy Bone

Marrow

OUTER BONE

INNER BONE

(Inside a bone)

The outer layer of the bone is called compact bone.

Compact bone is made of osteons and periosteum.

Osteons are layers of dense bone with a canal inside for blood vessels.

Do you see the red blood vessels inside the osteons?

Osteon

Compact Bone

Sound it Out

1. OS
2. TEE
3. ON

Sound it Out

1. KOM
2. PAKT
1. BONE

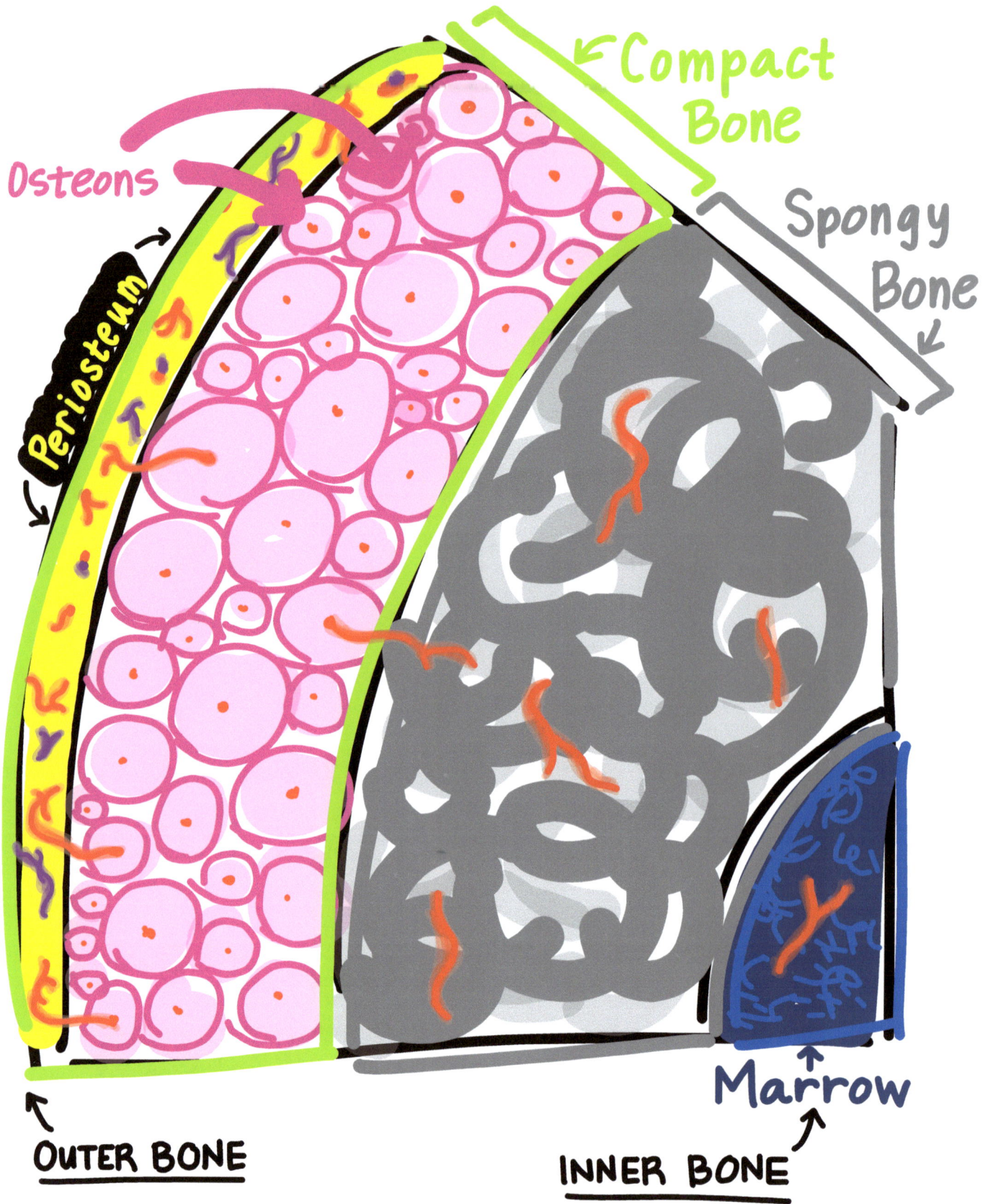

Osteons

Periosteum

Compact Bone

Spongy Bone

Marrow

OUTER BONE

INNER BONE

(Inside a bone)

The inner layer (the middle) of the bone is made of marrow and spongy bone.

Marrow is a soft substance that fills the spongy bone and makes blood cells.

What are osteons?
What is compact bone made of?
What is periosteum?

Spongy Bone

Marrow

Sound it Out
1. MER
2. O

Sound it Out
1. SPUN
2. JEE
1. BONE

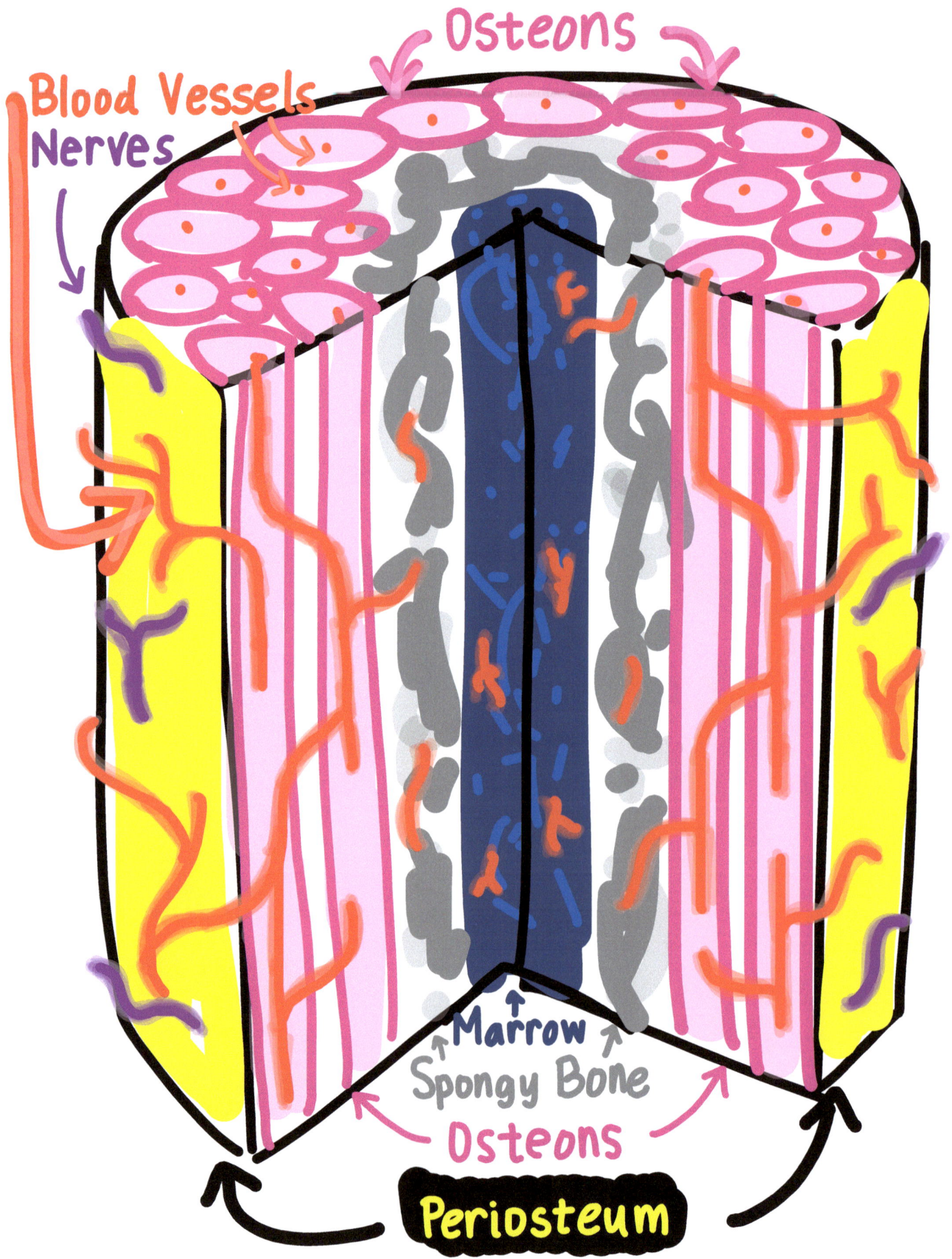

Osteons

Blood Vessels
Nerves

Marrow
Spongy Bone
Osteons

Periosteum

Starting from the middle of the bone going outward:

Marrow

Spongy Bone

Osteons

Periosteum

Blood vessels go from the outer membrane of the periosteum into the bone!

This is how your bone heals when it is fractured or broken.

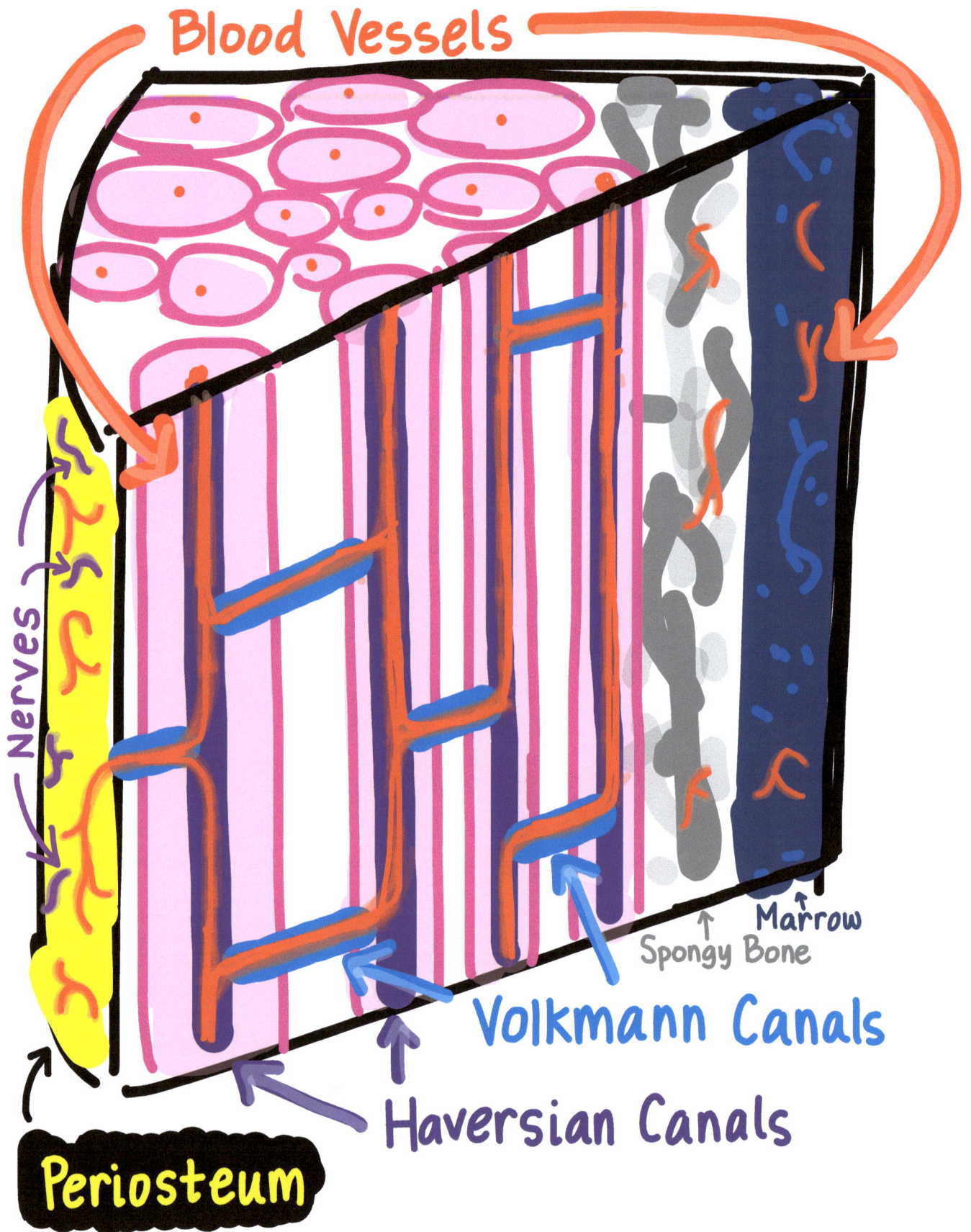

Blood Vessels

Nerves

Periosteum

Volkmann Canals

Haversian Canals

Spongy Bone

Marrow

Blood vessels are located
in the canals of the osteons.

The canals inside the osteons are called Haversian Canals. They go *up and down*.

Haversian Canal

Sound it Out

1. HUH
2. VER
3. SHUN

1. KUH
2. NAL

Volkmann Canal

Sound it Out

1. VOLK
2. MUN

1. KUH
2. NAL

The blood vessels move from *side to side* between osteons through Volkmann canals.

Volkmann canals also connect the outer periosteum to the inner osteons.

The outer compact bone is made of periosteum and osteons.

The inner bone is made of marrow inside spongy bone.

Blood vessels go from the outer periosteum to the osteons through volkmann canals that go from *side to side*.

Blood vessels move *up and down* the osteons through haversian canals.

Blood vessels inside the bone are the way fractured or broken bones heal.

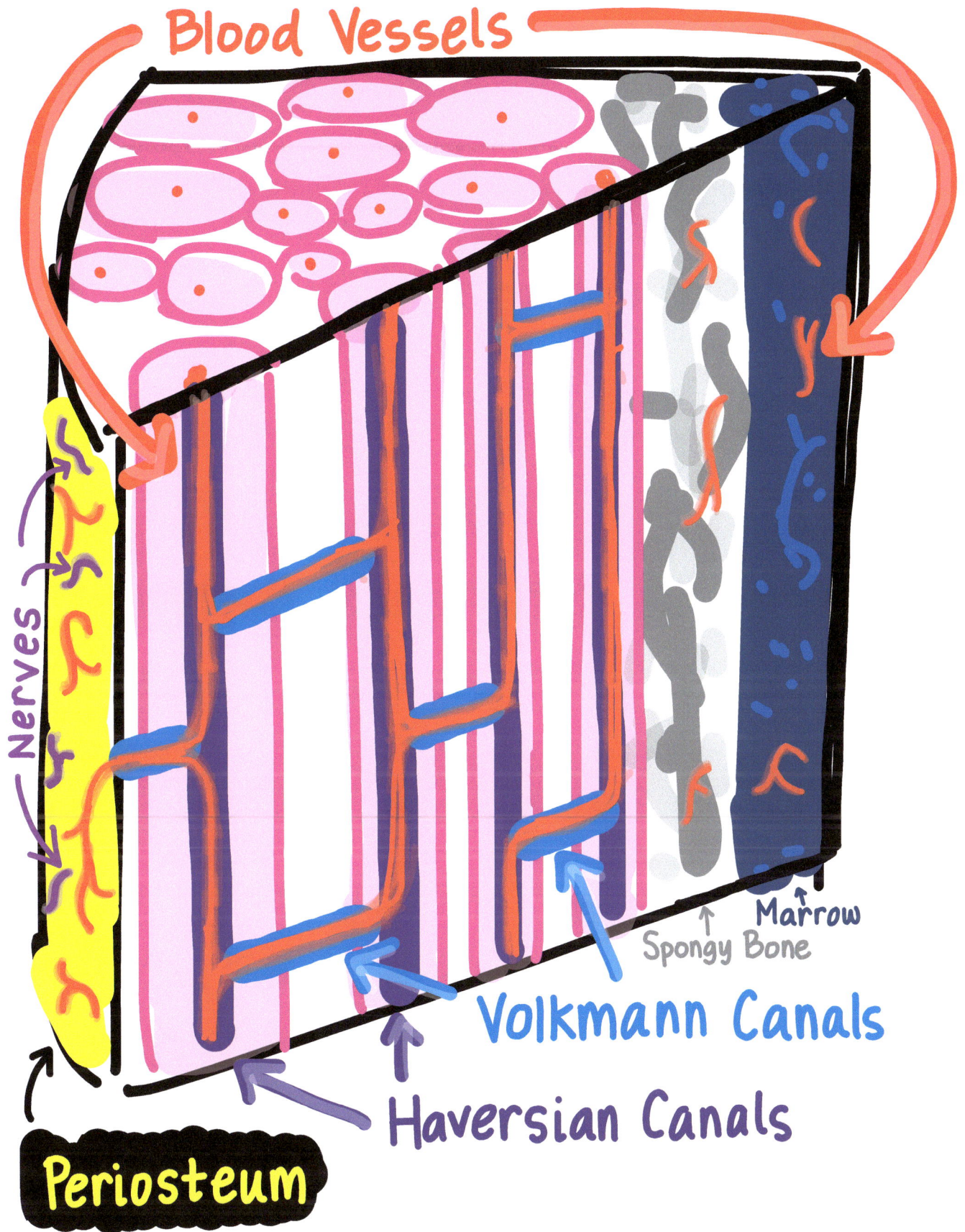

Blood Vessels

Nerves

Periosteum

Haversian Canals

Volkmann Canals

Spongy Bone

Marrow

What is the periosteum?

What is the difference between the haversian canals and the volkmann canals?

What is compact bone?

What is the difference between the epiphysis and diaphysis?

CONGRATULATIONS!

You are now a bone anatomy expert!

MUSCLE

There are 3 types of muscle in your body:

Skeletal
SKEL-EH-TUL

Cardiac
KAR-DEE-AK

Smooth
SMOOTH

Skeletal muscles are <u>voluntary</u> muscles. This means that <u>you</u> <u>control</u> the movement.

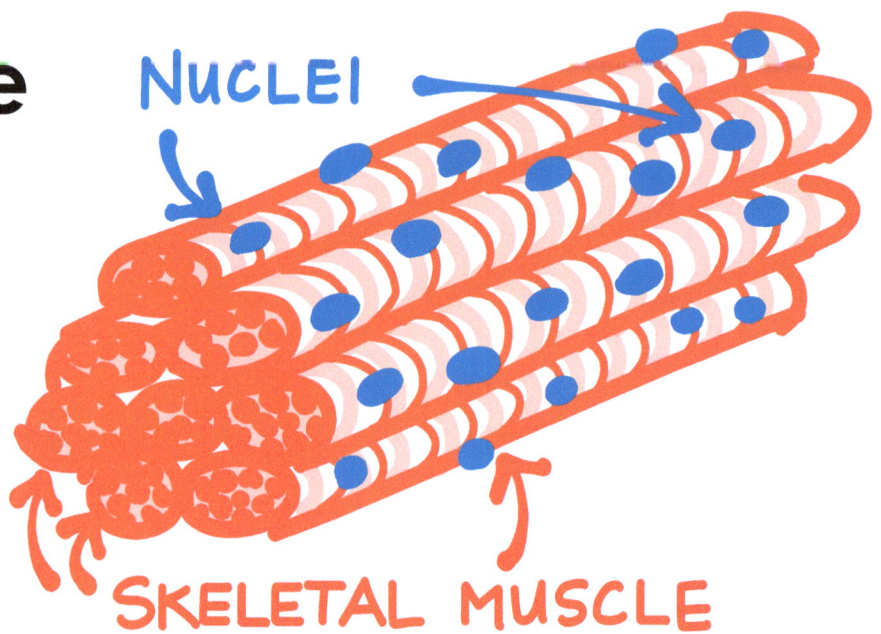

NUCLEI

SKELETAL MUSCLE

Skeletal muscles are striated. Striated means striped.

Do you see the striated skeletal muscle?

Skeletal muscles work together with the skeleton to move your body.

Striated

Sound it Out

1. STRI
2. A
3. TED

Skeletal muscles are attached to bones by tendons.

Tendons are made of white fiber-like tissue.

Tendon

Sound it Out
1. TEN
2. DUN

Tendons are VERY strong. They have to be because they hold the skeleton together.

How many bones are in a human skeleton?

Tendons allow you to run, walk, jump, lift and dance.

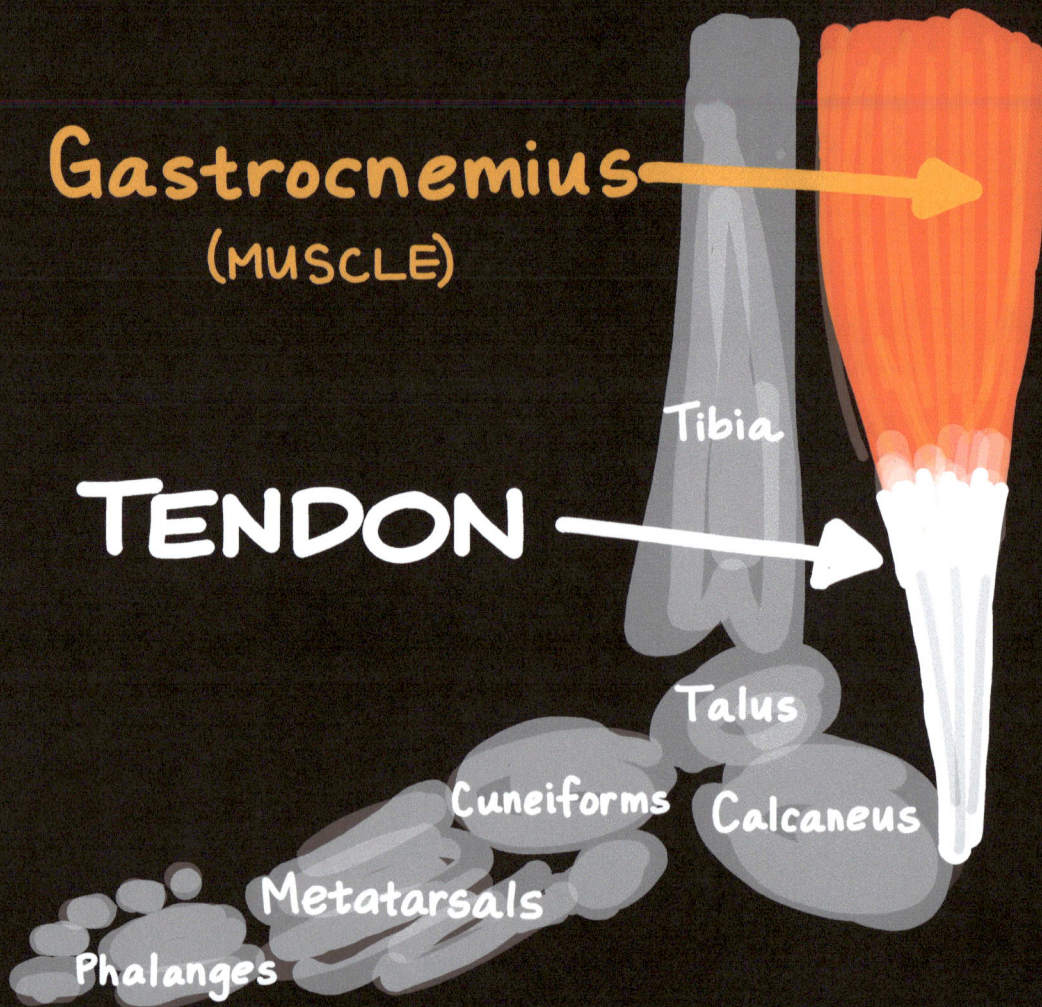

Gastrocnemius
(MUSCLE)

Tibia

TENDON

Talus

Cuneiforms

Calcaneus

Metatarsals

Phalanges

Gastrocnemius

This is a **tendon** that connects the heel to the **gastrocnemius** muscle so you can run and jump!

Sound it Out

1. GAS
2. TROK
3. NEE
4. MEE
5. US

Say the names of the bones in the lower leg and foot.

Tibia = TIB-EE-UH

Talus = TAL-US

Cuneiforms = Q-NEE-UH-FORMS

Calcaneus = KAL-KAN-EE-US

Metatarsals = MET-UH-TAR-SULS

Phalanges = FUH-LAN-JEES

The **tendon** connects
the **calcaneus bone** to the
gastrocnemius (skeletal) **muscle.**

Tendons connect muscle to bone.

Ligaments connect bone to bone.

Ligaments connect the femur (thigh bone) to the tibia (shin bone) at the knee joint.

Cartilage is in your joints, ears, nose and throat!

Ligament

Sound it Out

1. LIG
2. UH
3. MENT

Cartilage

Sound it Out

1. KAR
2. TEH
3. LEJ

The **patella** is the knee cap.

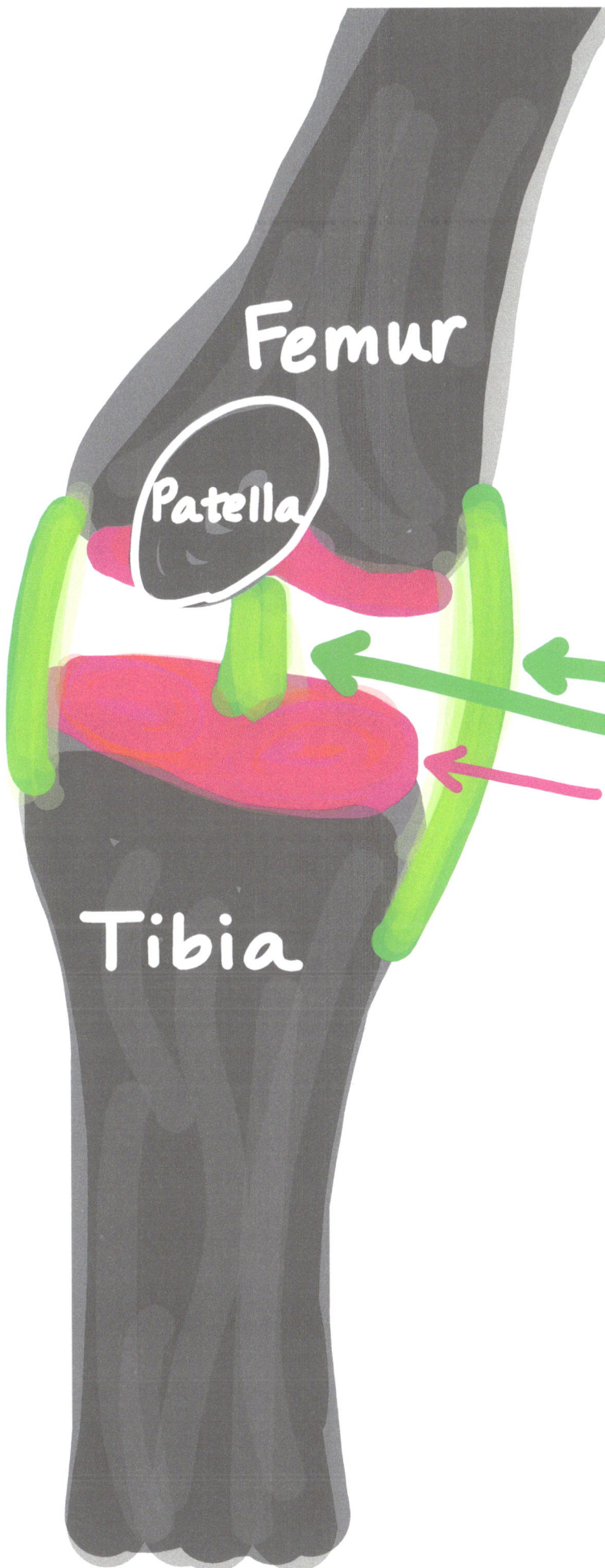

Femur

Patella

Tibia

LIGAMENT

Cartilage

Cartilage provides cushion for this knee joint.

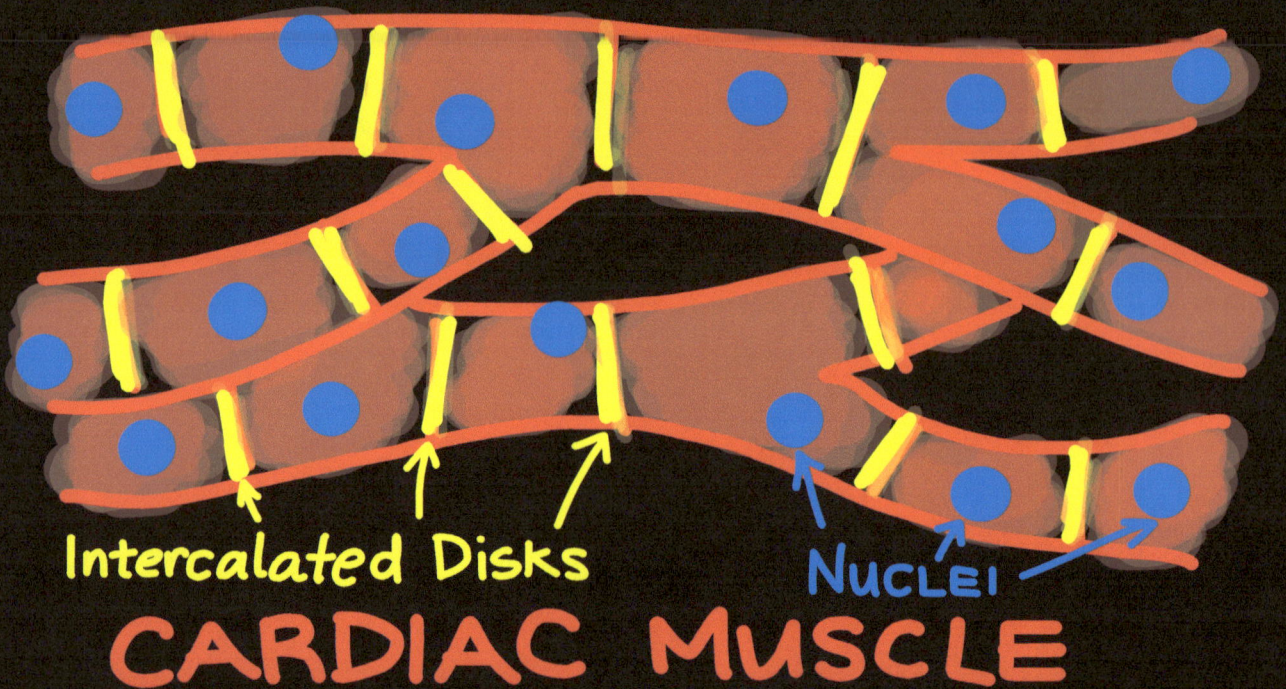

Intercalated Disks   NUCLEI

CARDIAC MUSCLE

Cardiac muscle is <u>involuntary</u>.

<u>You do not control</u>
your heart beat.

Cardiac muscle does not get tired. Cardiac muscle contracts to push blood out of the heart and relaxes to draw blood into the heart.

joined at **intercalated disks.**

Intercalated disks create the **striations** in cardiac muscle.

Intercalated   **Disks**

Sound it Out

1. IN
2. TER
3. KUH
4. LA
5. TED

Sound it Out

1. DSKS

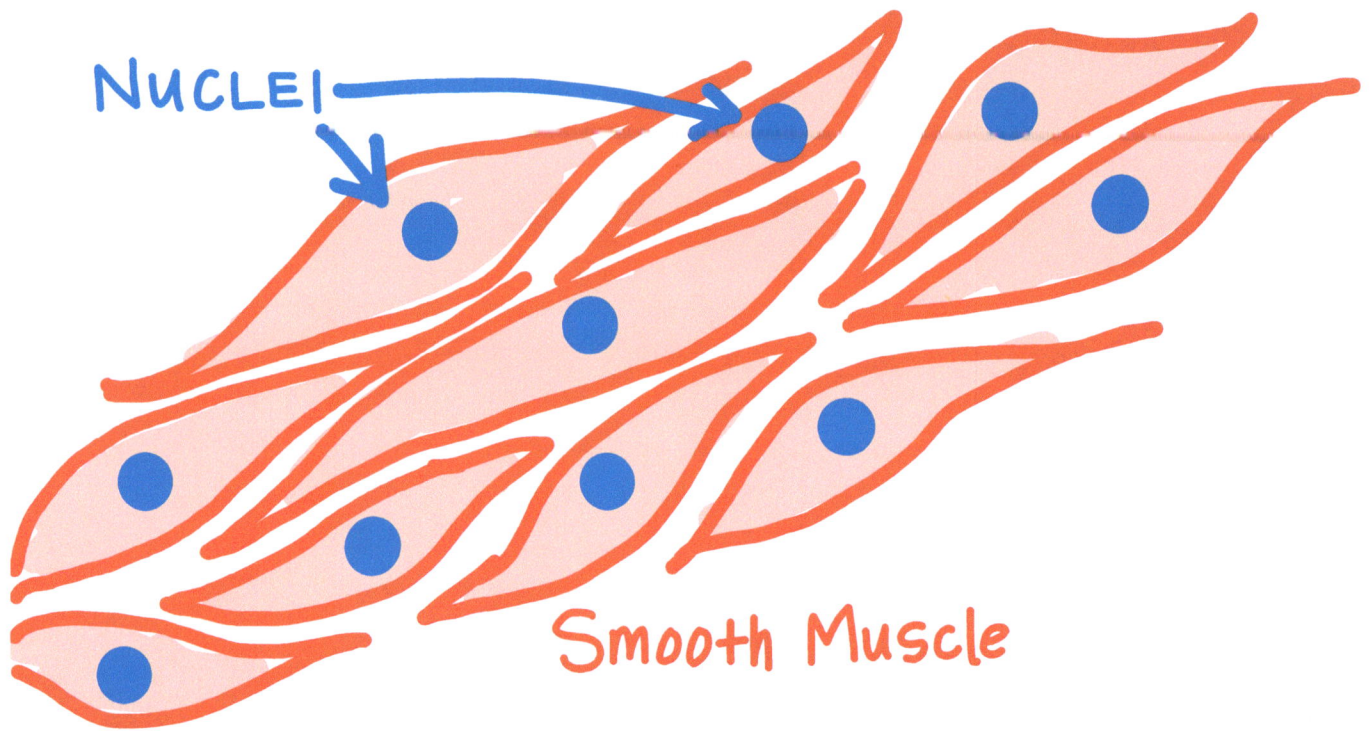

NUCLEI

Smooth Muscle

Smooth muscle is <u>involuntary</u>, it works automatically in your body.

Smooth muscle is in your stomach, intestines and even in your eye!

REVIEW:

Skeletal muscle is <u>voluntary</u>, striated and it is found all over your body.

Skeletal muscle connects to bones through tendons, like the calcaneus to the gastrocnemius.

Ligaments connect bone to bone, like the femur to the tibia.

Cardiac muscle is <u>involuntary</u> and found only in the heart.

Intercalated disks make the striations in cardiac muscle.

Smooth muscle is <u>involuntary</u> and found in your eye.

We know what bones and muscles look like, we know their anatomy. But how do our bones and muscles move?

When the muscle moves, the bones move because bones are connected to muscle through tendons.

What makes our muscles move?

The brain!

Neurons carry messages from our brain to our muscles.

(Review the structure and function of the neuron in Neurology: The Amazing Central Nervous System - book 3 of the Series - and elements in Chemistry: The Atom and Elements - book 2 of the Series)

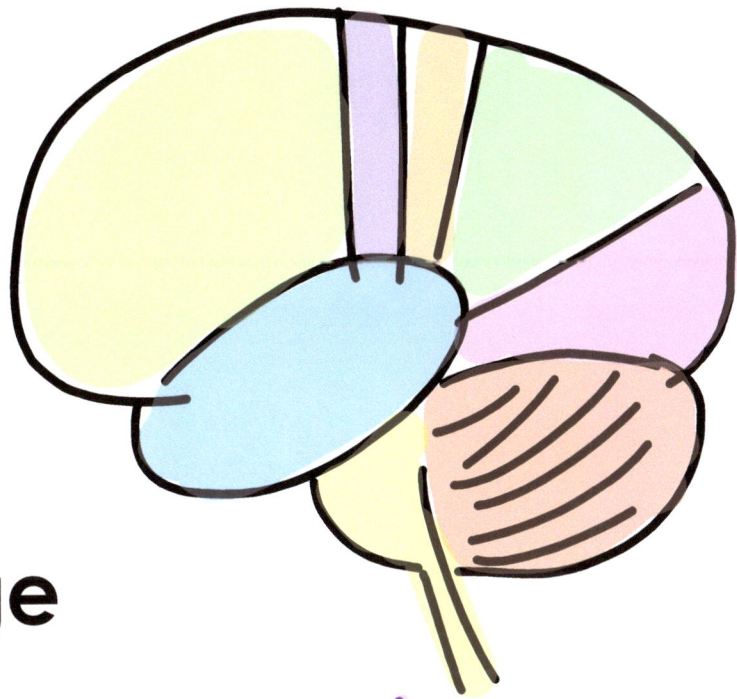

The message
to move is sent
from the brain,
through the
neurons, to the
muscle.

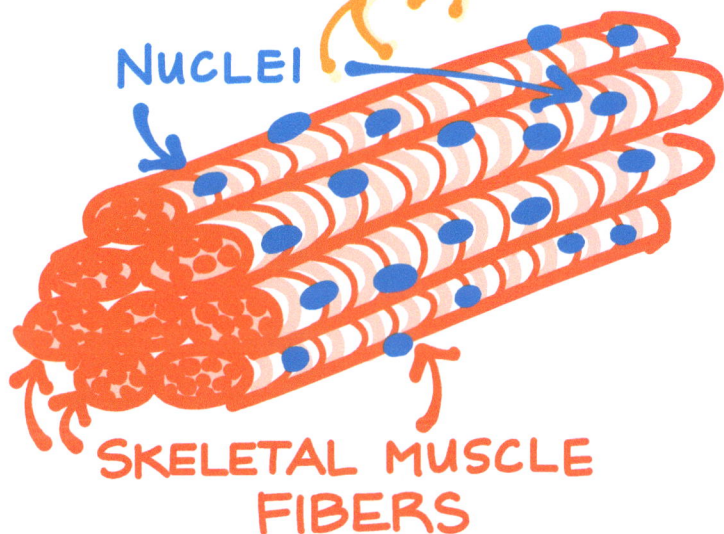

NUCLEI

SKELETAL MUSCLE
FIBERS

Axon Terminal

We are going to look at how the message is sent from the neuron to the muscle.

This happens at the Neuromuscular Junction.

Sound it Out
1. NUR
2. O
3. MUS
4. KU
5. LER

Sound it Out
1. JUNK
2. SHUN

(A junction is a meeting place)

The neuromuscular junction involves this part of the neuron, called the axon terminal, communicating with the muscle fibers.

Some super cool terms to know before we begin:

Calcium (Ca^{2+})
KAL-SEE-UM

Acetylcholine (ACh)
UH-SEE-TiL-KO-LEEN

Receptor
REE-SEP-TER

Synaptic vesicle
SIN-AP-TIK   VES-EH-KL

Sarcolemma
SAR-KO-LEM-UH

Neuromuscular Junction

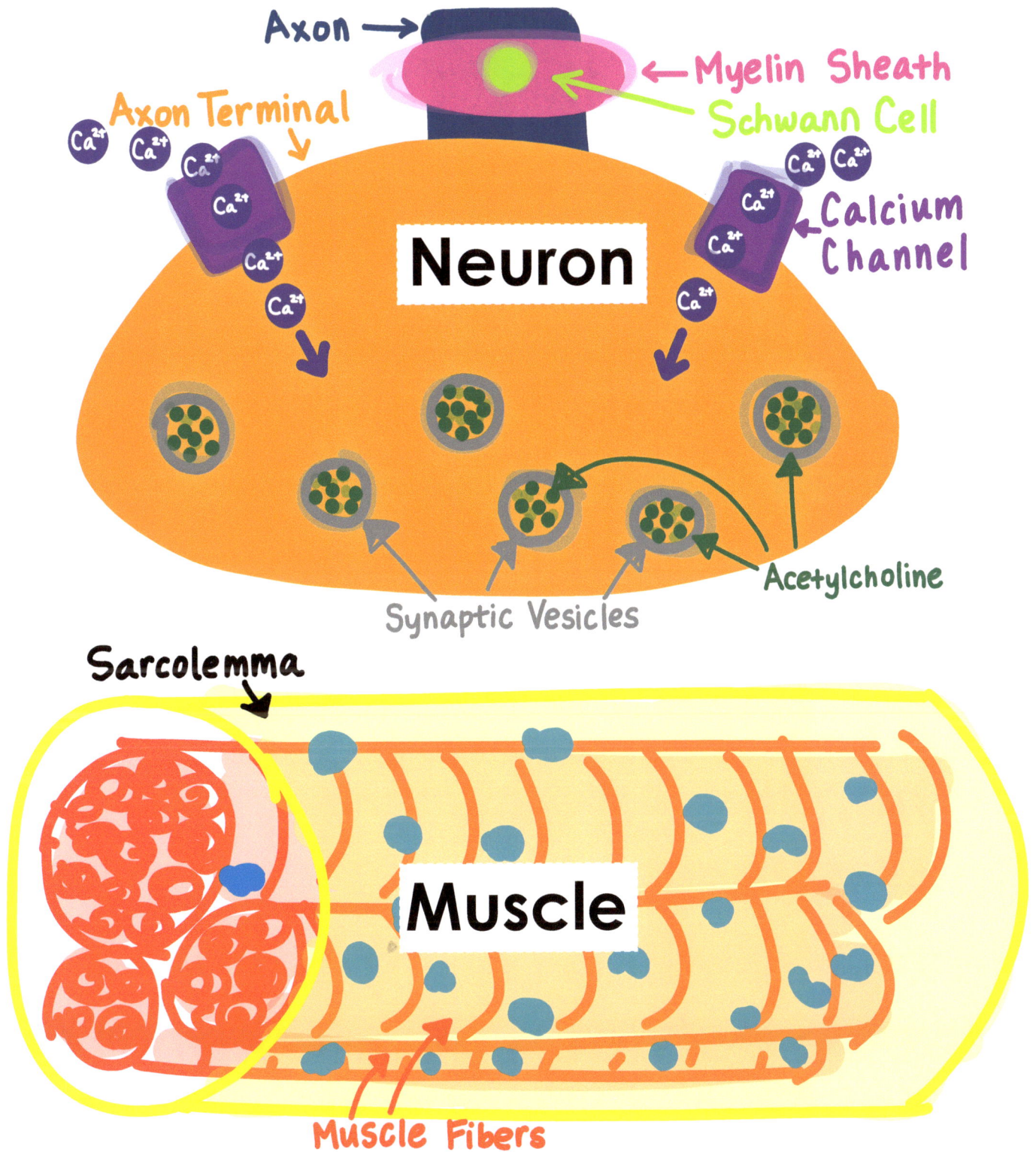

Axon →

← Myelin Sheath

Schwann Cell

Axon Terminal

Ca²⁺ Ca²⁺ Ca²⁺ Ca²⁺ Ca²⁺ Ca²⁺

Ca²⁺ Ca²⁺

← Calcium Channel

Neuron

Ca²⁺

Acetylcholine

Synaptic Vesicles

Sarcolemma

Muscle

Muscle Fibers

When the message reaches the axon terminal, it allows the element calcium (Ca^{2+}) to enter into the axon terminal.

The entry of calcium (Ca^{2+}) into the axon terminal causes the synaptic vesicles to come out of the axon terminal.

When the synaptic vesicles open, they release acetylcholine (ACh) into the space between the neuron and the sarcolemma (the membrane around the muscle fibers).

Neuromuscular Junction

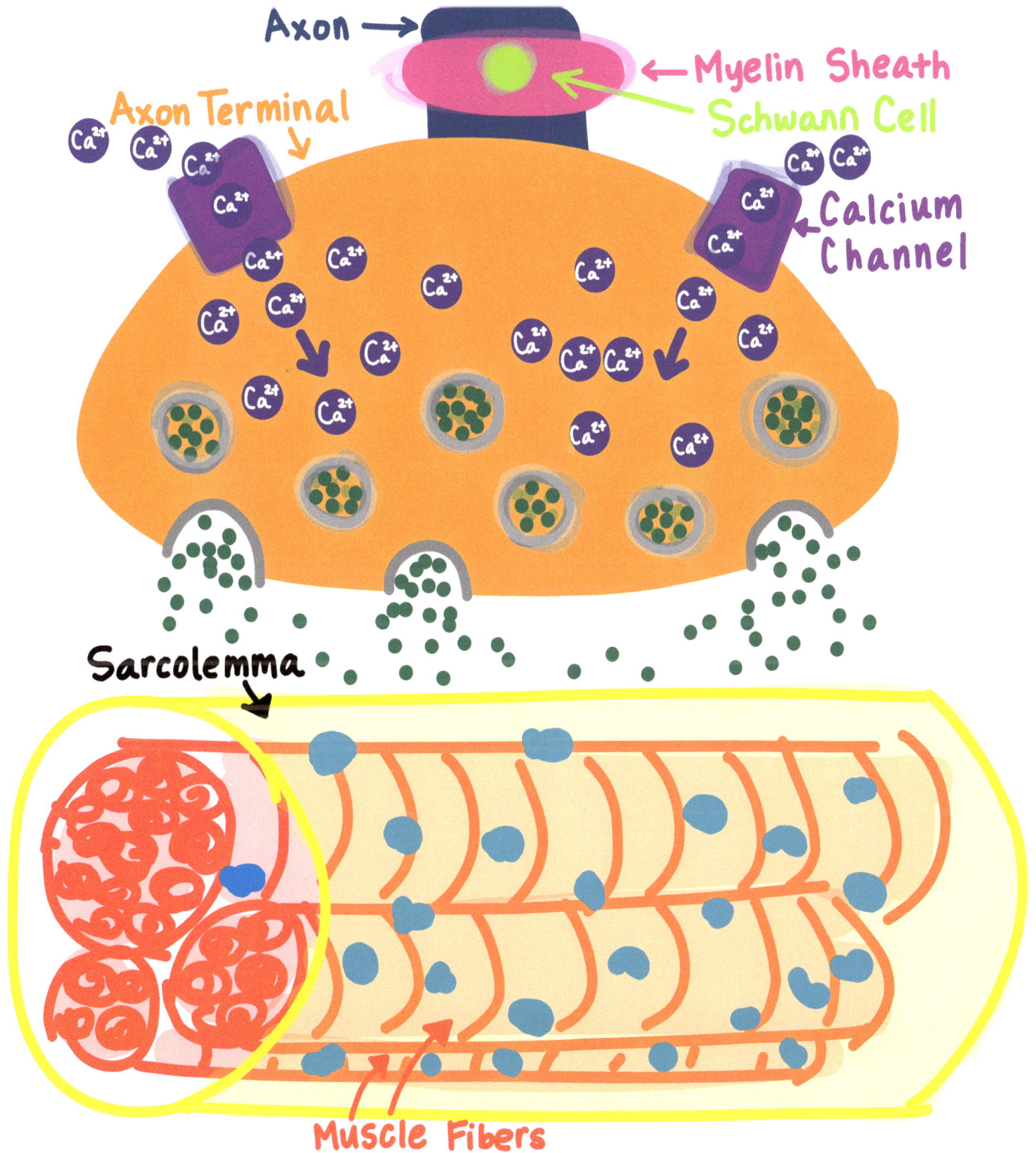

Axon →

← Myelin Sheath

Schwann Cell

Axon Terminal

Ca²⁺ Ca²⁺ Ca²⁺ Ca²⁺

Ca²⁺ Ca²⁺

Calcium Channel

Ca²⁺ Ca²⁺ Ca²⁺ Ca²⁺ Ca²⁺ Ca²⁺ Ca²⁺ Ca²⁺ Ca²⁺ Ca²⁺ Ca²⁺ Ca²⁺ Ca²⁺ Ca²⁺ Ca²⁺ Ca²⁺ Ca²⁺ Ca²⁺

Sarcolemma

Muscle Fibers

Next, acetylcholine (ACh) binds to the ACh receptors on the sarcolemma.

Review:

The message allows calcium into the axon terminal which causes the synaptic vesicles to release acetylcholine into the space between the neuron and the sarcolemma.
Then, ACh binds to the ACh receptors on the sarcolemma.

What is the sarcolemma?
What is the name of this junction where the neuron and muscle fiber meet?

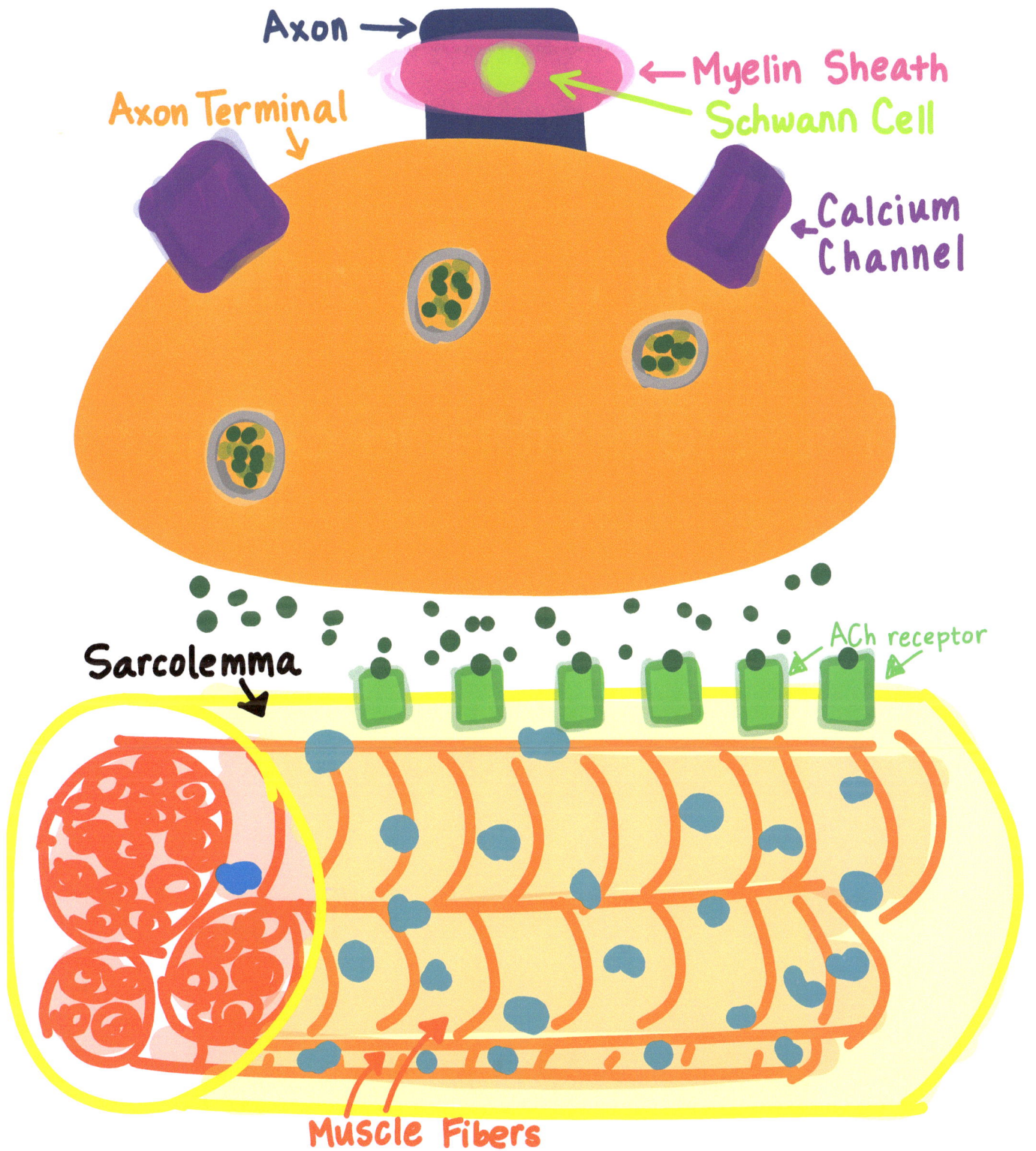

Axon

Myelin Sheath
Schwann Cell

Axon Terminal

Calcium
Channel

ACh receptor

Sarcolemma

Muscle Fibers

After ACh binds to the
ACh receptor, it allows
the element sodium (Na⁺)
to enter into the sarcolemma.

When enough sodium (Na⁺)
enters the sarcolemma,
the muscle is able to contract!

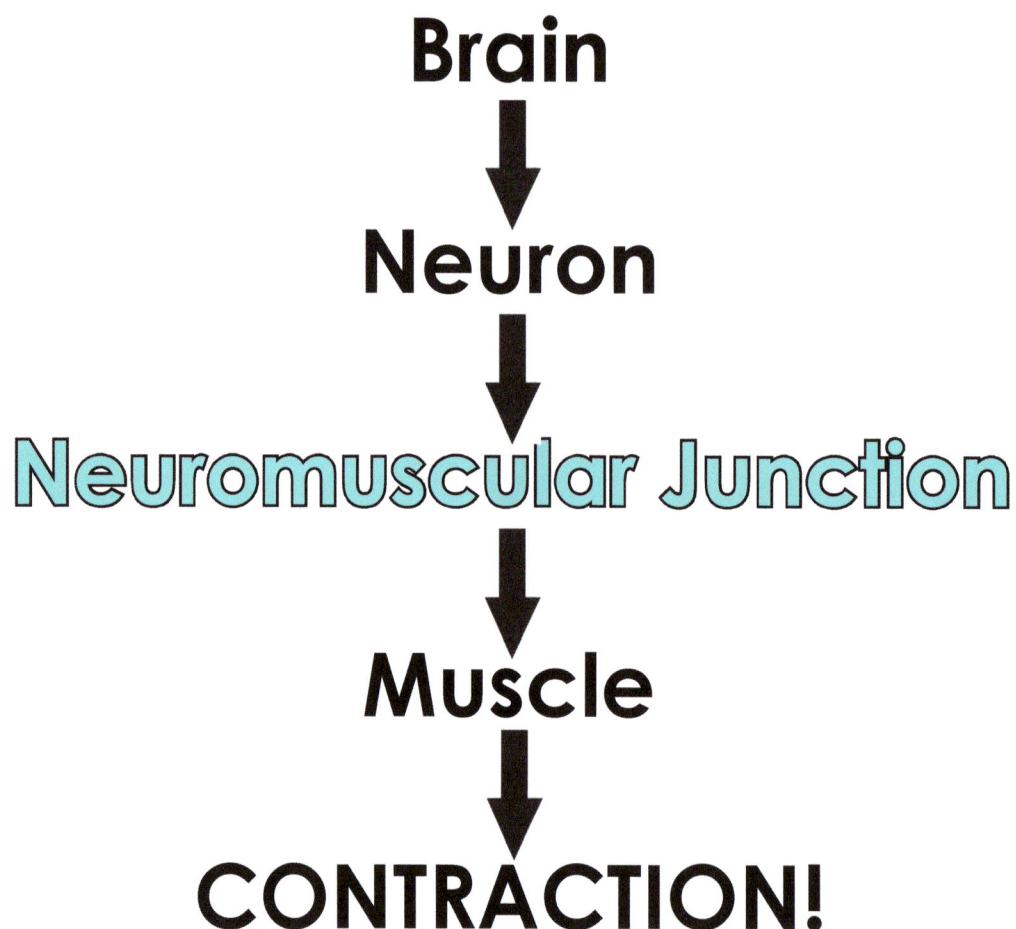

Brain

↓

Neuron

↓

Neuromuscular Junction

↓

Muscle

↓

CONTRACTION!

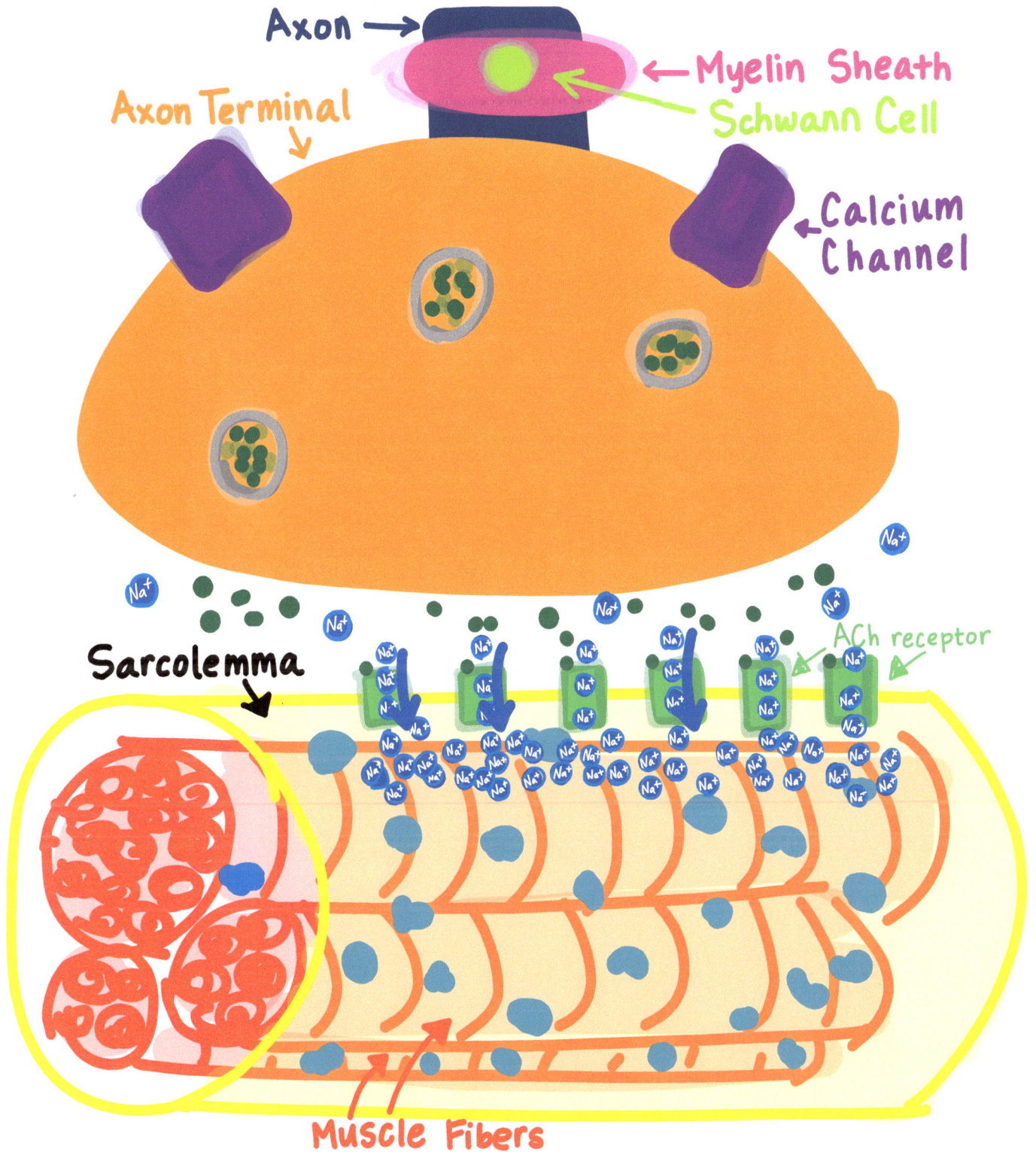

Axon

Myelin Sheath
Schwann Cell

Axon Terminal

Calcium
Channel

Na+

Na+

Na+

Na+

Na+

Na+

Na+

Na+

Na+

ACh receptor

Sarcolemma

Na+ Na+ Na+ Na+ Na+ Na+ Na+ Na+ Na+ Na+ Na+ Na+ Na+ Na+ Na+ Na+ Na+ Na+ Na+ Na+ Na+ Na+ Na+ Na+ Na+ Na+

Muscle Fibers

Sodium enters the sarcolemma,
causing the muscle to contract.

A human adult has 206 bones.

The epiphysis is the end of a long bone, the diaphysis is the shaft of a long bone.

The outer compact bone is made of osteons and periosteum.

The inner bone is made of spongy bone and marrow.

Haversian canals go *up and down.* Volkmann canals go *side to side.*

Skeletal muscle is voluntary and striated. Skeletal muscles are attached to bones by tendons.

Cartilage provides structure and cushion in your joints and nose.

Ligaments connect bone to bone.

Cardiac muscle is involuntary and striated from intercalated disks. (ONLY in the heart).

Smooth muscle is involuntary and automatic. (Stomach, eyes).

Movement in our body is started by a message from the brain that travels through neurons to our muscles. The message causes calcium (Ca^{2+}) to enter into the axon terminal, which signals acetylcholine to be released from the synaptic vesicles of the neuron and bind to the ACh receptors on the sarcolemma of the muscle, which allows sodium (Na^+) to enter and cause contraction of the muscle.

New Vocabulary!

Anatomy

Physiology

Bone

Skeleton

Epiphysis

Diaphysis

Periosteum

Compact Bone

Osteon

Marrow

Spongy Bone

Haversian Canal

Volkmann Canal

Skeletal muscle

Cardiac muscle

Smooth muscle

Striated

Tendon

Gastrocnemius

Cartilage

Tibia

Talus

Cuneiforms

Calcaneus

Metatarsals

Phalanges

Ligament

Patella

Voluntary

Involuntary

Intercalated Disk

Neuromuscular Junction

Axon Terminal

Calcium (Ca^{2+})

Acetylcholine (ACh)

Receptor

Synaptic Vesicle

Sarcolemma

Sodium (Na$^+$)

You are a bone, muscle, and neurology expert!

The Super Smart Science Series for ages 0-100:

#1 - Cellular Biology
Organelles, Structure, Function

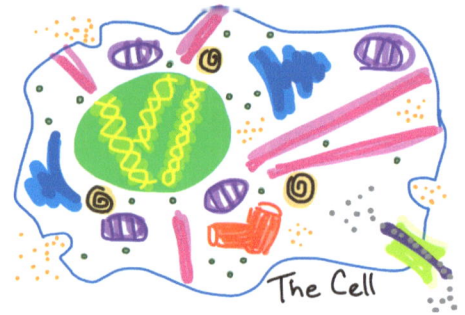

The Cell

#2 - Chemistry
The Atom and Elements

#3 - Neurology
The Amazing Central Nervous System

#4 - Astronomy
The Solar System

#5 - Anatomy & Physiology #1
Bones, Muscles, and The Stuff That Connects Bones and Muscles

#6 - Anatomy & Physiology #2
Body Systems

#7 - Cardiology
The Incredible Heart

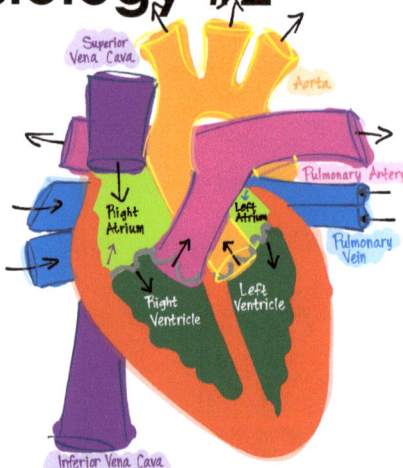

...and more!
www.SuperSmartScienceSeries.com

Draw YOUR bones, muscles, ligaments, tendons, neurons and label the parts!

www.ingramcontent.com/pod-product-compliance
Lightning Source LLC
Chambersburg PA
CBHW042001100426
42813CB00019B/2949